Paulo Roberto Queiroz
Mateus Costa
Laura Lisieux

Association between mutations in the KRT5, KRT14 and COL7A1 genes in two types of cancer

Paulo Roberto Queiroz
Mateus Costa
Laura Lisieux

Association between mutations in the KRT5, KRT14 and COL7A1 genes in two types of cancer

Molecular analysis of genes

ScienciaScripts

Cover image: www.ingimage.com

This book is a translation from the original published under ISBN 978-3-330-75859-9.

Publisher:
Sciencia Scripts
is a trademark of
Dodo Books Indian Ocean Ltd. and OmniScriptum S.R.L publishing group

120 High Road, East Finchley, London, N2 9ED, United Kingdom
Str. Armeneasca 28/1, office 1, Chisinau MD-2012, Republic of Moldova, Europe
Printed at: see last page
ISBN: 978-620-8-30250-4

Index

ASSOCIATION BETWEEN MUTATIONS IN THE *KRT5* AND *KRT14* GENES IN EPIDERMOLYSIS BULLOSA SIMPLEX

Mateus Costa De Santana * Paulo Roberto Martins Queiroz **

* Graduating student in the Biomedicine course at the University Center of Brasília - UniCEUB.

** Biologist. PhD in Animal Biology from the University of Brasília. pqsilva@uol.com.br

Introduction

Epidermolysis Bullosa (EB) is one of a group of genetic dermatological diseases characterized by the formation of blisters on the skin due to defects in structural proteins, which are important in the adhesion of the epidermal and dermal-epidermal layers. This defect favors the appearance of blisters when the tissue suffers external aggression and even high temperature variations (ALMEIDA JR, 2002).

Epidermolysis Bullosa is a rare disease that affects people of different ethnicities and there is no distinction between males and females. Several studies have been carried out on different populations and it has been shown that the United States has the most accurate record of the disease. Data collected from 1986 to 2002 allowed us to correlate a prevalence rate of 8: 1,000,000 people in 1990, with around 3,300 cases of EB being discovered, and the incidence standing at 1: 1,000,000 live births between 1986 and 1990. It has been shown that the Simple EB group has the highest number of cases and the Localized EB subtype has the highest rate. The highest prevalence of Junctional EB cases was the milder Nao-Herlitz form. The most common recessive dystrophic form was Nao- Hallopeau Siemens. The prevalence and incidence should be similar among the other populations due to a small difference in the values compared to the United States registry, thus demonstrating that there is no distinction between geographic locations and ethnicities (BOEIRA, 2012).

The condition was classified into four main types following the meeting of the International Consensus on Epidermolysis Bullosa in 2008: EB Simple (EBS); EB Junctional (EBJ); EB Dystrophic (EBD); and Kindler Syndrome. This division is due to the number of different genes and clinical phenotypes of Epidermolysis Bullosa. EBS is divided into two main types: Suprabasal which occurs in layers above the basal keratinocytes of the epidermis, and basal which affects the basal keratinocytes. The most studied EBS genes are *KRT5* and *KRT14*, which are responsible for synthesizing keratins-5 and -14 of the intermediate filaments and are of autosomal dominant inheritance (FINE et al., 2014). Some studies have reported cases of EBS with autosomal recessive inheritance, but mainly in the *KRT14* gene, which represents only 5% of all EBS mutations and the main related factor is inbreeding (GARCiA et al., 2011).

The EBJ and EBD types do not show problems with dermo-epidermal adhesions in the basement membrane zone, but do occur in the lamina lucida and lamina densa, respectively (ALMEIDA JR, 2002). The EBJ genes are: *LAMA3*; *LAMB3*; *LAMC2*; *COL17A1*; *ITGA6*;

ITGA3; *and ITGB4.* In EBD, only the *COL7A1* gene has been reported *and,* in these forms, both autosomal dominant and recessive inheritance patterns are found among the subtypes (FINE et al., 2014; GHR, 2015a; GHR, 2015b).

Kindler Syndrome is a patterned dermatosis of autosomal recessive inheritance and the formation of blisters can occur at multiple levels in the intralaminar lucid and sublaminar dense basement membrane and presents a particular feature of photosensitivity. The gene associated with this syndrome is *FERMTl* (*KIND1*) (FINE, 2010).

The *KRT5* and *KRT14* genes together synthesize cytoskeletal proteins called keratin intermediate filaments. The proteins of the cytoskeleton are important for muscle contraction, cell movement, transportation of intracellular components, as well as determining the morphology of cells. The intermediate filament is present in all cells, providing strength and resistance to external damage. Its structure can be made of keratin, vimentin or desmin. However, they are present in different tissues (JUNQUEIRA; CARNEIRO, 2004; ALMEIDA JR, 2004).

The human body is protected by epithelial barriers which are important components of innate immunity, including the skin and the internal epithelial linings of the gastrointestinal and respiratory tracts. Because they are in direct contact with the external environment, the physical barriers prevent the invasion of pathogens, protect the epithelium against external aggression, prevent dehydration, produce peptides that act as antibiotics and epithelial lymphocytes that help fight microorganisms (ABBAS; LICHTMAN, 2009). Currently, there is not much detailed information on the structural changes of these proteins with EB. Knowledge of the abnormal changes helps to develop effective methods and treatments and thus improve the quality of life of patients living with this disease.

The aim of this study was to relate the phenotypes of Simple EB of the basal region based on the mutations found in the genes coding for the keratin-5 and -14 proteins.

Methodology

An experimental study was carried out using bioinformatics programs to elucidate the abnormal protein structures of keratin-5 and keratin-14, which represent the main subtypes of Epidermolysis Bullosa Simplex (EBS). This study was supported by a bibliographic review of national and international scientific journals related to EBS and the *KRT5* and *KRT14* genes.

National and international scientific articles and theses were searched in the *EBSCOhost*, Scielo, PubMed, Medline, Elsevier, Bireme and Google Scholar databases from 2000 to 2016 and classic articles were added to elucidate certain terms. The keywords were: "Epidermolysis Bullosa Simplex"; "Intermediate Filaments"; "Structure"; " *KRT5* "; and " *KRT14* " in English or Portuguese separately and combinations of them using the AND connective.

Bioinformatics

"Bioinformatics refers to the use of computational tools in the study of biological problems and issues, also covering applications related to human health such as the planning of new drugs" (VERLI, 2014).

The tools used in this work are available on the following websites: National Center for Biotechnology Information (NCBI) (http://www.ncbi.nlm.nih.gov/); The Sequence Manipulation Suite (SMS) (http://www.bioinformatics.org/sms2/); Protein Data Bank (PDB) (http://www.rcsb.org/pdb/home/home); and SWISS-MODEL (http://swissmodel.expasy.org/).

The NCBI website is a nucleotide database which, in addition to providing gene sequences from various species, offers the use of various tools to work with these sequences. First, a gene search was carried out in the NCBI nucleotide database to acquire information on the nucleotide sequences of *KRT5* and *KRT14*. This information was used to build a model of the protein, but before modeling, the SMS GenBank Trans Extractor tool was used to obtain the peptide sequences. These sequences were then analyzed using the Modelling tool on the SWISS-MODEL website, resulting in 3 models with sequences closest to the model sequence. The only sequences that represented keratin-5 and keratin-14 were the 2B helical domains of both proteins. The protein call numbers: 3tnu.1.A (Keratin-14) and 3tnu.1.B (Keratin-5) are deposited in the PDB protein database and were used to model the mutated proteins. A circular phylogenetic tree was made for the *KRT5* and *KRT14* genes using the NCBI BLAST tool, using the sequence NM_000424.3 sequence of the *KRT5* gene as a base, for the alignment between other sequences of the *KRT5* gene (M21389.1; NM_000424.2; BC024292.1) and for the base sequence NM_000526.4 of the *KRT14* gene with other *KRT14* sequences (BT007186.1; BC094830.1; NM_000526.3). Only 3 gene sequences were chosen, all of which had an identity above 95%. When constructing the graphs, it was necessary to use information from the human intermediate filament database (HIFD) to create a criterion for selecting mutations based on the most frequent nucleotide substitutions in the KRT5 and KRT14 genes, and the main domain with the highest mutation index in type I and type II keratins.

Results and Discussion

Epidermolysis bullosa simplex (EBS) of the basal type has three subtypes, these variations are the result of mutations occurring within specific regions of the *KRT5* or *KRT14* genes. Blisters are formed by the rupture of the basal cells present in the epidermal layer; these are cells called keratinocytes, which are responsible for synthesizing important keratins for the tissue, and these proteins are used to form the intermediate keratin filaments. The blisters represent a consequence of the weakening of the keratin proteins and, due to the lysis of these cells, some authors call this pathology Epidermolytic (ALMEIDA JR, 2002).

The clinical picture of each of the subtypes of BSS is classified by the way in which they manifest themselves. The main factors in differentiating them are where the blisters form, how intense they are and at what point in life the signs and symptoms begin to develop. Table 1 shows these three main features of the clinical picture, as well as other important findings such as anemia due to blood loss caused by the blisters breaking and other findings that have been observed in some patients (LIOVIC et al., 2004). The subtypes of basal-type EBS are: Localized EBS (formerly Weber-Cockaine); Intermediate Generalized EBS (formerly Koebner); and Severe Generalized EBS (formerly Dowling-Meara). The Localized form presents restricted blisters on the hands and feet, both on the back and on the palmar region and soles of the feet, but they are present in areas where there is a lot of physical friction. The two generalized forms can occur anywhere on the body and manifest from birth, but the intensity of the signs and symptoms between them are different. The Severe Generalized form shows hyperkeratosis on the palms and soles of the feet, and the presence of blisters on the mucosa in some patients (MACHADO et al., 2002; JERABKOVA et al., 2010). Figure 1 shows patients with the three subtypes of EBS.

Table 1: Clinical characteristics of the EBS subtypes.

	EBS-Located	**Severe generalized EBS**	**EBS-generalized intermediate**
Old names	Weber-Cockayne	Dowling-Meara, Hepertiforme	Other generalized, Nao Dowling-Meara, Koebner
Pattern of inheritance	AD	AD	AD
First symptoms	Early childhood	Birth	Birth
Distribution of Bubbles (Predominance)	Hands, palms, feet and soles.	Generalized	Generalized

Skin finding (frequency)*			
Bubbles	4+	4+	4+
Milia	Rare	1-2+	1+
Atrophic scars	Rare	2+	1+
Dystrophy / absence of wedge	Unusual	2+	1-2+
Hyperkeratosis	Focal (in some of the majority)	Generally diffuse	Focal
relative bubble concentration	Common	Common	Common
Extracutaneous involvement*			
Anemia	Absent	Variable	Absent
Growth retardation	Absent	Common	Absent
Dental	Little	Common	Variable
Gastrointestinal	Absent	2+ (cold)	Absent
Ophthalmology	Absent	Absent	Rare

Legend: AD - Autosomal Dominant; E BS - Simple Bullous Epidermolysis; * Frequency relative - absent; rare; 1+; 2+; 3+; 4+.

Source: Fine et al. (2014).

Figure 1: Examples of skin lesions in patients with EBS.

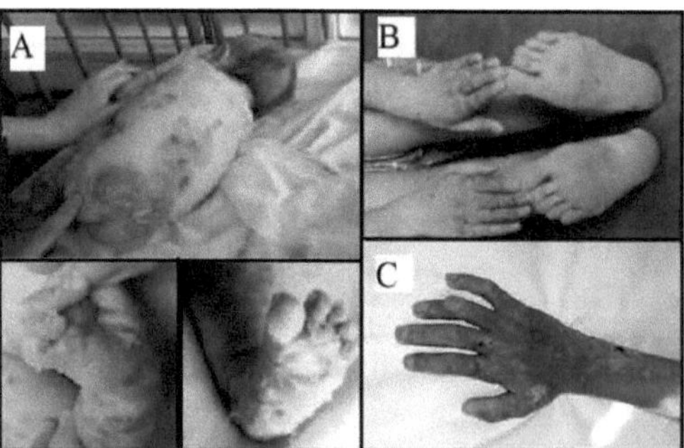

Caption: In **A**, a newborn with the severe Generalized form, the hand and foot have hyperkeratosis (BOEIRA, 2012); **B**, localized form; **C**, Hand of a child with the Intermediate Generalized form (TITEUX et al., 2006).

KRT5 and KRT14 genes responsible for encoding keratins -5 and -14

The *KRT5* and *KRT14* genes are responsible for synthesizing type II keratin 5 and type I keratin 14, respectively. They are a group of strong fibrous proteins that form the structural framework of some cells, such as those present in epithelial cells, nails, hair and other tissues. These genes belong to the keratin family, and there are two types of keratin, basic and acidic proteins (LANE; MCLEAN, 2004). The fibrous proteins come together to form intermediate filaments which can be synthesized by 70 different genes and have been separated into 6 families. The first family are the type I keratins (K9 to K23 and K25 to K28) found in soft tissues and because they have negative or neutral charges they are called acidic. The second is made up of type II keratins (K1 to K8 and K71 to K80) which are basic proteins that generate positive charges and are also found in soft tissues. Table 2 shows the six families of proteins that can make up the intermediate filaments, as well as the size and tissue in which they are synthesized (BRAGULLA; HOMBERGER, 2009).

Table 2: Description of the main protein families found in intermediate filaments.

TYPE	PROTEIN	SIZE (kD)	PLACE OF EXPRESSION
I	Acid keratins K9 - K23 and K25 - K28	40-60	Epithelial cells
II	Basic or neutral keratins K1 - K8 and K71 - K80	50-70	Epithelial cells
III	Vimentin	54	Fibroblasts/leukocytes/other cell types.
	Demine	53	Muscle cells
	Glial fibrillary acidic protein	51	Glia cells
	Peripherin	57	Peripheral neurons
IV	Neurofilament proteins		
	NF- Low molecular weight	67	Neurons
	NF- Medium molecular weight	150	Neurons

	NF-High molecular weight	200	Neurons
	α-Internexin	66	Neurons
V	Protein blades	60-75	Nuclear lamina of all cells
VI	Nestina	200	CNS stem cells

Legend: NF - Neurofilament; K - keratin; KD: kilodaltons; CNS - central nervous system.

Source: COOPER (2000); BRAGULLA; HOMBERGER (2014).

The isoelectric point (IP) (the point at which there is a balance between the positive and negative charge of a molecule in a solution) is responsible for this division. Even with the differences in IP, basic and acid keratins contain 30% of the same amino acids and sequences, and these are also observed in other intermediate filaments. As a result of these differences in charges, the intermediate filaments will have greater stability when compared to filaments made up of just one family (BRAGULLA; HOMBERGER, 2009).

More than 100 mutations have been found in the keratin-5 gene that cause at least one of the EBS subtypes and more than 60 mutations in the keratin-14 gene, these alterations being responsible for making the epidermis fragile and blistering easily when compared to people who have normal keratins. The *KRT5* and *KRT14* genes are located on chromosomes 12 and 17 (Figure 2) (GHR, 2016a; GHR, 2016b). Most mutations lead to amino acid substitutions, but what will influence the EBS subtype will be the disturbance caused by these changes during the assembly of the final protein. The *KRT5* gene is involved in 11 different pathologies and 10 diseases related to the *KRT14* gene (SZEVERENYI et al., 2008).

Figure 2: Cytogenetic localization of the *KRT5* and *KRT14* genes.

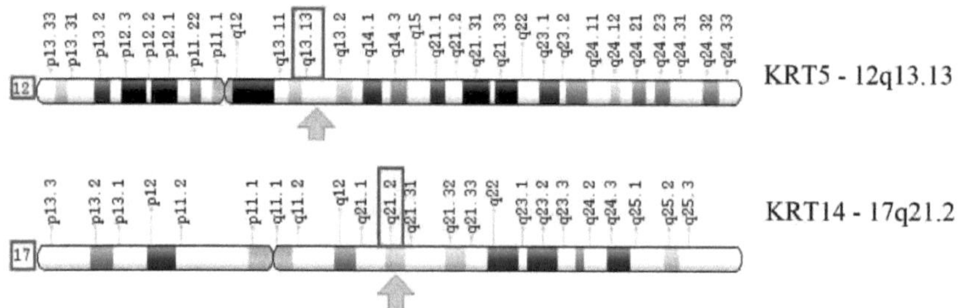

Source: GHR (2016A); GHR (2016b).

The main nucleotide changes caused by mutations in type I and II keratins are shown in figure 3. The main substitution is a transition from Adenine (A) to Guanine (G), the second transition changes Cytosine (C) to Thymine (T), and the third is a transversion from T to G. Transition-type substitutions are more frequent than transversions because the changes in the chemical structure of amino acids are less drastic than transversions. The factor behind these mutations in DNA is unknown, but the transition from C to T may be related to the deamination of 5-methylcytosine, which ends up becoming more susceptible to spontaneous deamination due to the instability of the CpG nucleotides and forming a thymine (SZEVERENYI et al., 2008).

Figure 3: Main nucleotide changes in type I and type II keratins.

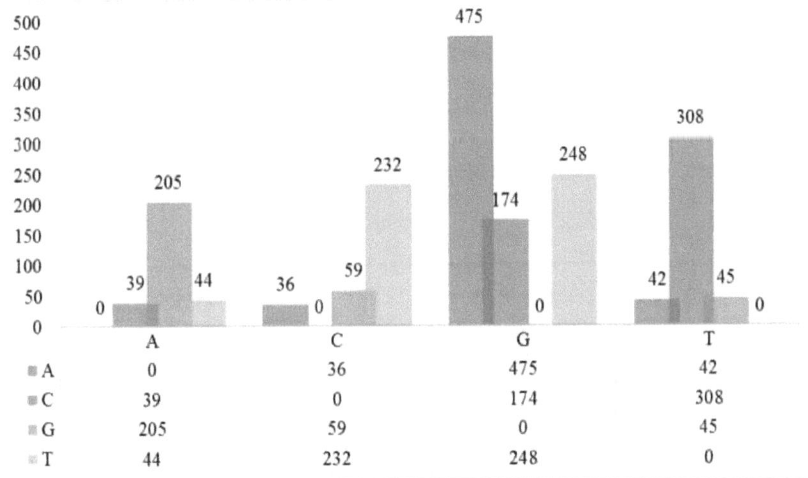

	A	C	G	T
A	0	36	475	42
C	39	0	174	308
G	205	59	0	45
T	44	232	248	0

Legend: Adenine (A); Cytosine (C), Guanine (G) and Thymine (T).

Source: Adapted from SZEVERENYI et al. (2008).

Keratins

Keratins (K) are proteins with an α-helix structure, as well as other proteins capable of forming intermediate filaments. The final structure of the protein is formed by the dimerization of the K5 and K14 monomers or by any of the proteins mentioned above in table 02, which are responsible for forming the intermediate keratin filaments. For this dimerization to occur, the centers of the α-helix present in the keratins need to interact, thus forming the K dimer, this central region later known as the superhelix. Next, the dimers join in parallel in opposite directions and maintain a distance between the C-Terminal and N-Terminal regions, forming the tetramer. After this stage, the protofilaments are assembled, which are tetramers linked by their C- and N-terminal ends. Then there is the junction of two protofilaments forming the protofibril and, finally, there is the interaction between four protofibrils forming the intermediate keratin filament (Figure 4) (JUNQUEIRA; CARNEIRO, 2004).

Figura 4: Assembly of the intermediate keratin filament.

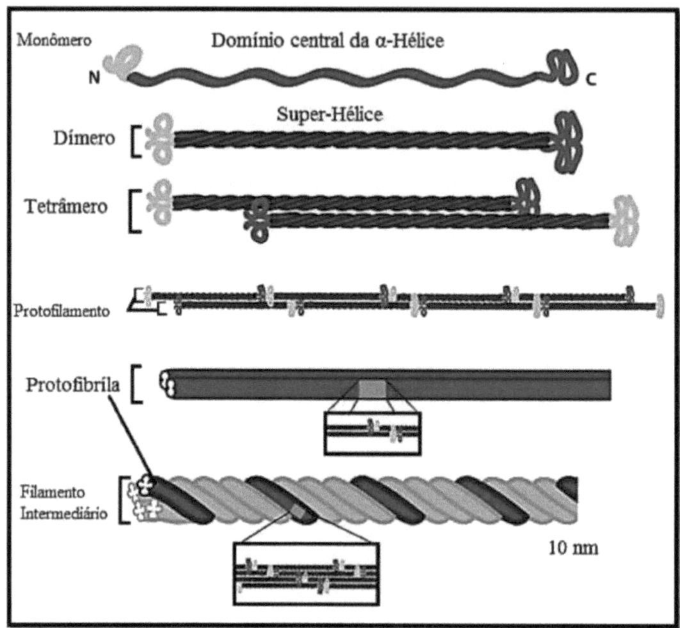

Legend: N - Amino-terminal portion; C - Carboxy-terminal portion.

Source: Adapted from PROTEiNAS (2016).

The interactions between acidic and basic keratins generate greater stability for the intermediate filaments when compared to those made up of homodimers, because the differences between the two types of keratins present a greater number of hydrophobic bonds, in addition to the electrostatic interactions of the keratins (LANE; MCLEAN, 2004).

The intermediate filaments are distributed within the cells, creating a network of connected filaments. To prevent these filaments from becoming loose in the cytoplasm, they are linked with proteins adhered to the cytoplasmic membrane. These proteins are the desmosomes and hemidesmosomes which will ensure adhesion between adjacent cells and with the basal lamina. When mutations affect the ends of the monomers, the bond between the tetramers and with the structures present in the cell membrane will be compromised.

and are therefore important factors in the development of pathologies (SZEVERENYI et al., 2008; ALMEIDA JR, 2004). Figure 5 shows the network of intermediate filaments, the location of the desmosomes and hemidesmosomes in the cell.

Figura 5: Distribution of the network of filaments, desmosomes and hemidesmosomes within the cell.

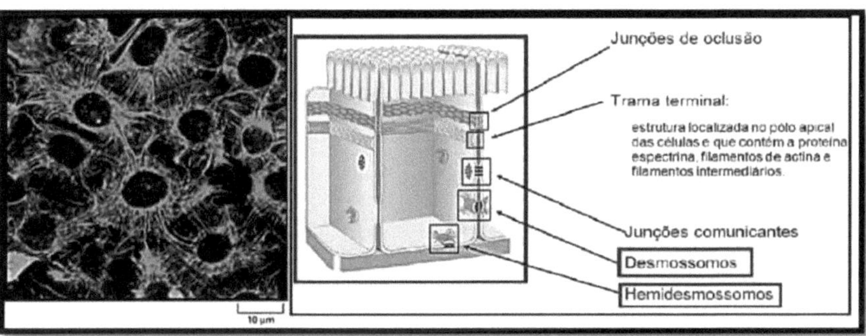

Legend: On the left, the network of filaments is shown in green by the immunofluorescence technique and, on the

on the right, the location of the desmosome and hemidesmosome structures.

Source: Adapted from COOPER (2000); JUNÇÔES (2016).

Keratin structures are divided into head, body and tail. The body of the intermediate filaments has a central α-helix domain (1A, 1B, 2A and 2B) which is separated by the non-helical binding segments (L1, L12 and L2). At the ends there are variable segments (V1 and V2), responsible for the variability observed in the structures of the intermediate filaments and the basic keratins have segments (H1 and H2), which are found between the variable segments and helical domains (Figure 6) (ALMEIDA JR, 2004). In the helical head and tail portion of the intermediate filaments, there are two conserved sites, the first is at the beginning of helical domain 1A (corresponding to 23 amino acids) and the second is the final 32 amino acids of helical segment 2B. These regions are considered critical because they represent the sites of interaction between the ends of the keratin heterodimers. Modifications in the conserved parts tend to express the most severe phenotype of EBS and are called *hotspots*, because they are points within the DNA molecule that are more likely to be mutated than other regions (JERÂBKOVÂ et al., 2010).

Figure 6: Structure of the keratin-5 and -14 proteins.

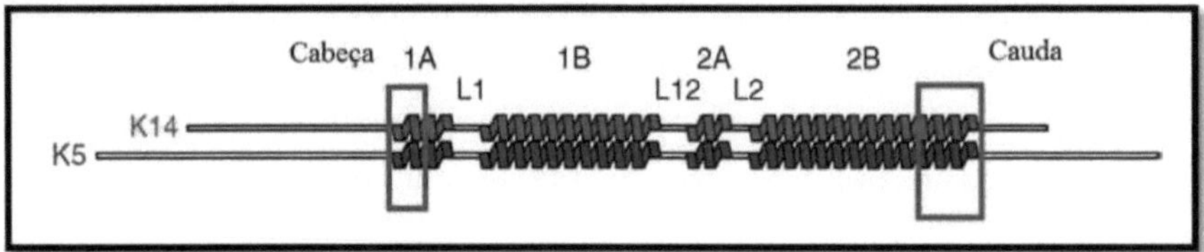

Legend: K5 - keratin 5; K14 - keratin 14; the marked regions represent the two conserved sites involving the two proteins.

Source: Adapted from LEE et al. (2012).

For α-helix structures to form, seven amino acids must be repeated (*hepta* repeats). This is important in the interactions between the helical domains of K5 and K14 for the superhelix to form. These repeats are represented as (**a** - **b** - **c** - **d** - **e** - **f** - **g**) in the first helix, and in the second as (**a'** - **b'** - **c'** - **d'** - **e'** - **F** - **g'**). The amino acids normally found in the **a** and **d** positions have apolar chemical characteristics and are located at the interface of the superhelix. The **e** and **g** residues are polar and directed towards the solvent side, responsible for increasing the specificity in the electrostatic interactions between the helices (LEE et al., 2012).

The formation of a superhelix between two α-helices normally follows the three principles proposed by the peptide Velcro hypothesis. Firstly, the residues (**a** and **d**) need to be hydrophobic, such as isoleucine, leucine or valine, since the amount of hydrophobic interactions increases stability and facilitates the process of dimerization of the superhelix. Secondly, residues (**e** and **g**) contain the amino acids glutamate and lysine respectively, which are responsible for ionic interactions between proteins. Thirdly, residues (**b,** **c** and **f**) need to be hydrophilic as they are in direct contact with the solvent and are responsible for assembling the α-helix due to intramolecular interactions, forming hydrogen bonds. Positions (**a** - **d** - **e** - **f**) are responsible for the intermolecular interactions that assemble the superhelix (Figure 7). Each *hepta* repeat corresponds to an α-helix with two turns (MASON; ARNDT, 2004).

Figura 7: Assembly pattern of keratin superhelixes showing Heptad repeats.

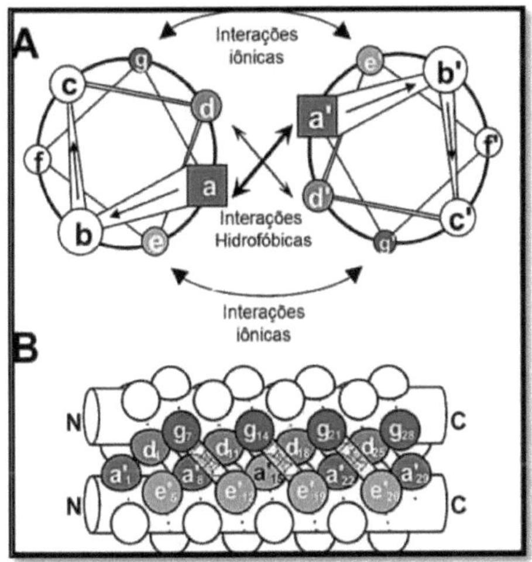

Legend: (A) bottom view of the protein - hydrophobic interactions between the protein residues (**a, a'**) and (**d, d'**), ionic interactions (**e, g'**). (B) side view of the protein - Colors, **a** (purple), **d** (blue), **e** (orange) and **g** (red), the green structure in the middle represents a polar residue responsible for specific bonds.

Source: Adapted from MASON; MÜLLER; ARNDT (2008).

Bioinformatics

In bioinformatics, the gene sequences used were acquired from the NCBI database and correspond to the call number NM_000424.3 for the *KRT5* gene and NM_000526.4 for the *KRT14* gene. Two phylogenetic trees were made, one for each gene, using NCBI's BLAST tool, in order to observe the distance between the chosen sequences and other sequences of the same genes deposited in NCBI, separately (Figure 8 and 9). The sequences of the chosen genes were also used to analyze the structures of the K5 and K14 proteins (ZHANG et al., 2000).

As no information was found on the main EBS mutations, the *Human Intermediate Filament Database* (HIFD) website was used as a basis for selecting some mutations. The criteria used to choose the mutations were: the main nucleotide and amino acid changes; the protein domain most affected; and the availability of a model in SWISS-MODEL (Figures 3, 10 and 11). The chosen mutations described in the *KRT5* gene were: nucleotide A1210G, amino acid K404Q; nucleotide C1283T, amino acid A428V; nucleotide 1285_1299del15, amino acid R429_A433del; and for the *KRT14* gene were: nucleotide C1228T, amino acid Q410X; nucleotide T1243C, amino acid Y415H (SZEVERENYI et al., 2008).

Figure 8: Alignment and circular phylogenetic tree of the *KRT5* gene.

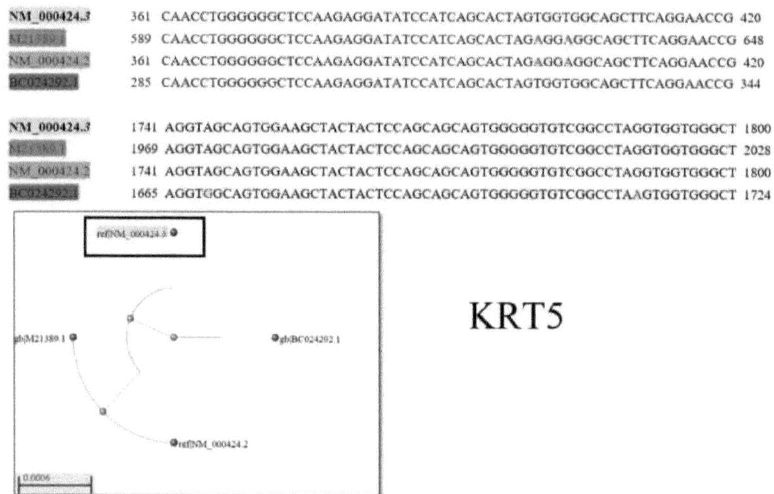

Source: Adapted from ZHANG et al. (2000).

Figure 9: Alignment and circular phylogenetic tree of the *KRT14* gene.

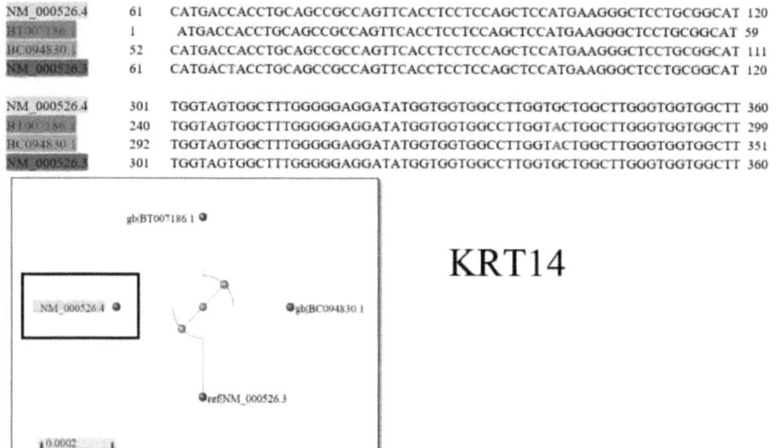

Source: Adapted from ZHANG et al. (2000).

The results of the phylogenetic trees were 0.0006 for the *KRT5* gene *and* 0.0002 for *KRT14* using the minimum evolutionary distance method, with a maximum distance of 0.75. To arrive at these values, the four gene sequences were aligned and the nucleotides that are different from the sequences that are marked were added together. The number of different nucleotides was then divided by the total number of nucleotides in the marked gene sequences. The result represents the length between the aligned sequences. Therefore, the sequences of the *KRT5* and *KRT14* genes that are marked have a distance of 0.0006 and 0.0002 to the closest point of the last branch. The events that caused these branches made the sequences NM_000424.3, NM_000526.4 and variants susceptible to mutations that cause EBS (SEIDEL et al., 2008).

Figures 10 and 11 show that most of the mutations are in the 1A and 2B helical domains and L12 in the non-helical domain. The main mutations in the *KRT5* gene occurred in the L12, 1A and 2B domains, causing the subtypes of EBS-Localized, Intermediate Generalized and Severe, respectively. While in the *KRT14* gene they were in domains 1A for Severe Generalized EBS and 2B for the subtypes of Localized and Intermediate Generalized EBS (SZEVERENYI et al., 2008).

Figure 10: Main domains affected in keratin 5.

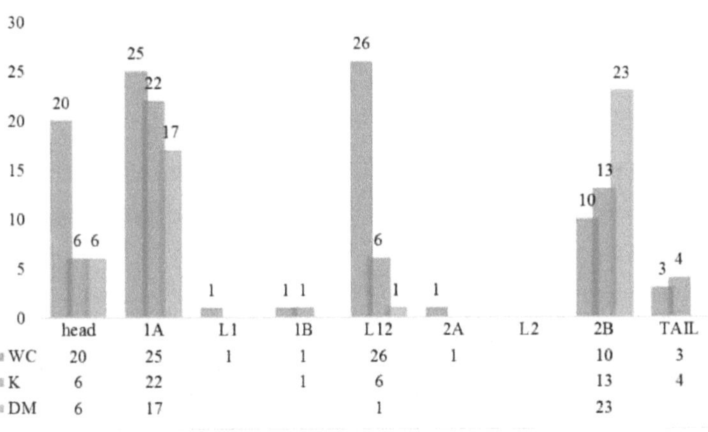

Legend: WC - Localized; K - Intermediate Generalized; DM - Severe Generalized.

Source: Adapted from SZEVERENYI et al. (2008).

Figure 11: Main domains affected in keratin 14.

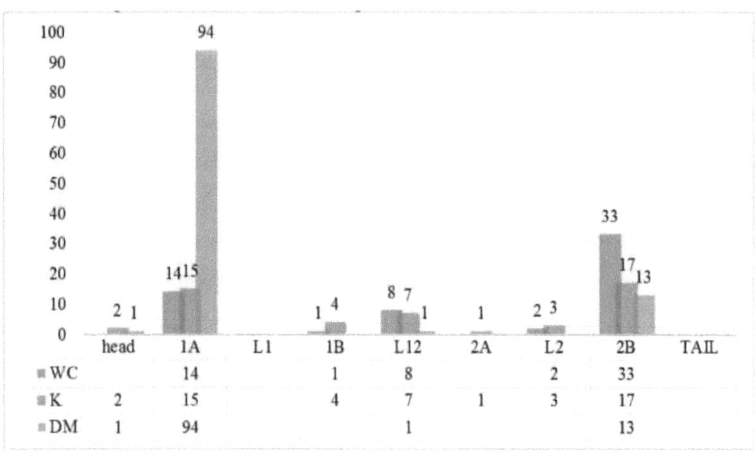

	head	1A	L1	1B	L12	2A	L2	2B	TAIL
WC		14		1	8		2	33	
K	2	15		4	7	1	3	17	
DM	1	94			1			13	

Legend: WC - Localized; K - Intermediate Generalized; DM - Severe Generalized.

Source: Adapted from SZEVERENYI et al. (2008).

Two different methods can be used to build three-dimensional protein structures: the first is by experimental approach, *Nuclear Magnetic Resonance* (NMR), *X-ray crystallography* or *cryo-electron microscopy*, and the results are then deposited in databases such as the PDB and available for viewing in specific programs. The advantage of this methodology is that it is highly accurate in predicting structures, but there are difficulties in obtaining enough samples for the experiment, poor quality results, the method is difficult to carry out and it is a time-consuming technique (PACHECO, 2009).

The second method is the computational approach, consisting of *homology modeling*, *Threading* and *Ab initio*. These methods use the sequences from the experimental approach, which have been deposited in the databases, in which they will be compared to a "problem" sequence. If the existing sequence has a similarity of 25% or more, homology modeling can be used, but when they are less than 25% the best way to analyze them is by sequence-structure compatibility, *Threading*, which is the choice of the best model or *Ab initio*, which uses quantum methods to elucidate the structure by emitting molecular waves generated by the distribution of atomic nuclei and electrons around these molecules (PACHECO, 2009; SANTOS FILHO; ALENCASTRO, 2003). The elucidation of the normal protein will help to understand the alterations caused by mutations and thus analyze the possible changes or loss of protein function (DIAS; TAVEIRA, 2011).

Molecular Modeling by Homology

Homology modeling is based on molecular evolution through the divergence of genes according to duplication and association with mutations and, from these modifications, data is generated on the formation of gene families. Proteins will have homologous regions, conserved domains and highly variable regions. The changes in homologous proteins can be identical, similar or dissimilar, but even with these differences, the three-dimensional structure of the proteins is maintained, because the ancestral form is important for carrying out the function normally (SANTOS FILHO; ALENCASTRO, 2003). In homology modeling, a template is used to compare with the problem sequence and predict a three-dimensional structure with a degree of similarity between the two molecules. This will guarantee the fidelity of the structure and the level of similarity between the two, but when looking for an abnormal variation in the protein, the sequence can be modified and the effect of this difference observed (SOUZA, 2008).

The α-helix 2B domains of keratin-5 and -14 were constructed by homology modeling in the SWISS-MODEL program, in which the sequences were simulated with and without changes. The changes in this domain were linked to the development of EBS (BIASINI et al., 2014).

The 2B domain of keratin-5 starts at positions 350 to 477 and contains 131 amino acids. Based on this information, changes were made to the peptide sequence representing the three mutations chosen. The first two mutations substituted a lysine for a glutamine at position 404 and the other substituted an alanine for a valine at position 428 in the protein's 2B domain, but did not change the protein. In the third mutation, there was a deletion from position 429 to position 433, and the formation of a pleated beta-sheet was observed, which was normally represented by an α-helix (Figure 12). The 2B domain of keratin-14 starts at position 295 up to position 422 and 2 different mutations were inserted into two separate sequences, the first mutation was represented by a stop codon at position 410, normally occupied by glutamine and thus decreasing the size of the 2B domain. The second mutation replaced the tyrosine at position 415 with a histidine, but did not show a structural change (Figure 13) (ARNOLD et al., 2006; SZEVERENYI et al., 2008).

Amino acid changes or deletions in the 2B domain were noticeable in some mutations, but those that did not change the structural conformation of the protein have the ability to modify intermolecular interactions, and the dimerization of keratins is highly dependent on these interactions. Even small changes in the domains can cause significant alterations to develop some pathology (RUGG et al., 2007).

Figure 12: Comparison of the different mutations occurring in the 2B region of keratin-5.

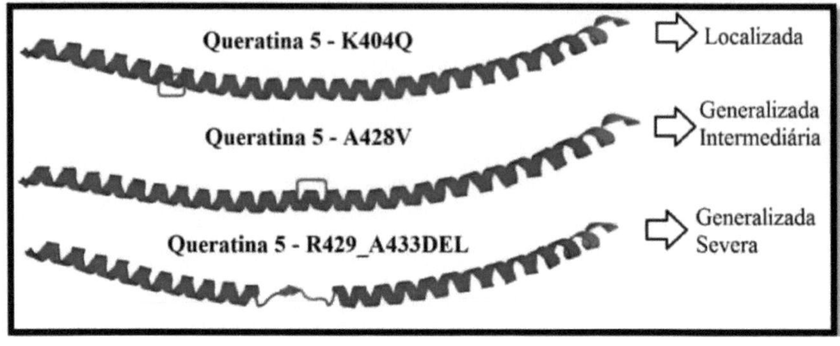

Legend: Model built using the SWISS-MODEL program.

Source: BIASINI et al. (2014).

Figure 13: Comparison of the effect of two mutations occurring in the 2B region of keratin-14.

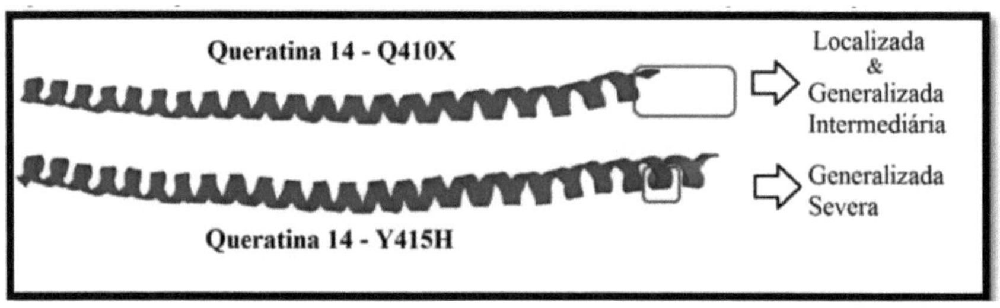

Legend: Model built using the SWISS-MODEL program.

Source: BIASINI et al. (2014).

Association between genotype and phenotype

The results of the mutations chosen for K5 show the following phenotypes: the first mutation, L404Q, represented the Localized subtype; the A428V mutation, the Intermediate Generalized form; and in the third mutation there was a deletion of five amino acids from position 429 to 433, which was represented by the Severe Generalized form. The phenotypes of the two mutations in K4 were: The first mutation represented two phenotypes, the Localized subtype and the intermediate generalized form, this mutation at position 1228 changed the glutamine codon to a stop codon, which interrupted protein synthesis from position 410; the last mutation, Y415H, led to the more severe phenotype of EBS (Severe Generalized) (SZEVERENYI et al., 2008). Table 3 represents a list of possible mutations to be used in SWISS-MODEL, and in bold are the mutations chosen for the bioinformatics experiment.

Table 3: List of the main mutations described in the *KRT5* and *KRT14* genes and the correlated phenotypes.

Gene	Nucleotide	Change	Amino-acid	Change	Domain	Subtype	Authors
KRT5	A1210G	Replacement	**K404Q**	Replacement	2B	Localized	Schuilenga-Hut et al.(2003).
	C1283T	Replacement	**A428V**	Replacement	2B	Generalized I	Rugg et al. (2007).
	1285_1299del15	Deletion	**R429_A433del**	Deletion	2B	Generalized S	Kemp et al. (2005).
	T605A	Replacement	L202Q	Replacement	1A	Localized	Gao et al. (2015).
	T932G	Replacement	L311R	Replacement	1B	Localized	Ciubotaru et al. (2003
	C1054A	Replacement	R352S	Replacement	2A	Localized	Yasukawa et al. (2006
	G508A	Replacement	E170K	Replacement	1A	Generalized I	Oldak et al. (2011).
	G794C	Replacement	R265P	Replacement	1B	Generalized I	Kang et al. (2010).
	G506C	Replacement	R169P	Replacement	1A	Generalized S	Müller et al. (2006).
KRT14	C1228T	Replacement	**Q410X**	Replacement	2B	Generalized I; Localized	Arin et al. (2010).
	T1243C	Replacement	**Y415H**	Replacement	2B	Generalized S	Arin et al. (2010).
	?	Unknown	M119I	Unknown	1A	Localized	Chen et al. (1995).
	740_748delCCTACCTGAinsGAA	Indel	A247_K250delinsE	Indel	1B	Localized R; Localized	Bolling et al. (2011).
	A346G	Replacement	L116E	Replacement	1A	Generalized I	Rugg et al. (2007).
	G863T	Replacement	R288L	Replacement	2A	Generalized I	Minakawa et al. (2013
	G348C	Replacement	K116N	Replacement	1A	Generalized S	S0rensen et al. (1999)

Key: Del - Deletion; Ins - Insertion; Generalized I - Intermediate; Generalized S - Severe; A - adenine; C - cytosine; T - thymine; G - Guanine; R - arginine.

Source: Adapted from SZEVERENYI et al. (2008).

It has been observed that mutations in the *KRT5 and KRT14* genes have different pathogenic structural effects between the subtypes and that these modified proteins alter the metabolism of keratinocytes, but it is not clear how these alterations occur in cellular and tissue terms. Some studies have shown that the expression of some genes has been altered due to mutations in the K5 and K14 protein genes. In 2011, a study was carried out in Canada with the aim of observing the gene expression profile of 6 patients diagnosed with EBS and selecting 6 healthy people with the same characteristics (height, weight and age approximately 3 years apart) to be the control for each patient. Two patients had the subtype (Severe Generalized EBS) and four patients had the milder subtype (Localized EBS). Firstly, a difference in the expression of 28 genes was observed in patients with EBS (counting both subtypes) when compared to control subjects. The differences in the expression profile between patients (EBS-Generaliza Severa or EBS-Localized) and control subjects were 41 and 6 genes, respectively. Changes in the expression of 17 genes were observed between patients with EBS and Severe Generalized EBS and only one gene with EBS and EBS-localized patients (Figure 14) (BCHETNIA et al., 2011).

Figure 14: Study of the change in expression caused by different mutations in the *KRT5* and *KRT14* genes.

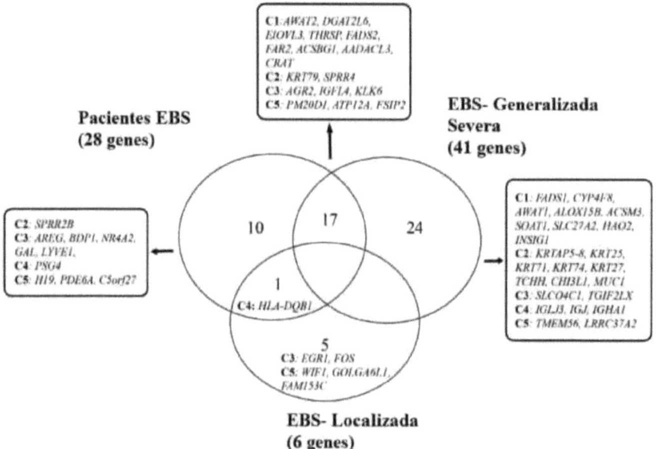

Caption: EBS - Epidermolysis Bullosa Simplex; Cl - Genes involved in lipid biosynthesis; C2 - Genes involved in keratinization; C3 - Genes involved in cell growth and proliferation; C4 - Genes involved in the immune response; C5 - Other functions.

Source: BCHETNIA et al. (2011).

The importance of elucidating the three-dimensional structure of these proteins helps in the development of intelligent drugs, allowing other programs to observe possible interactions between the target protein and the therapeutic drug, as well as minimizing side effects or maximizing cell repair mechanisms. The creation of new computer algorithms (a set of instructions for solving a problem) can help to develop new diagnostic methods or improve known methods. In order to develop specific drugs for EBS, it is necessary to discover a molecule capable of interacting with the protein and improving its stability, preventing cell lysis. Another way of improving the treatment of the pathology would be the ability of the drug to control the expression of genes caused by mutations in the *KRT5* and *KRT14* genes (OLIVA, 2008).

Final considerations

Epidermolysis bullosa simplex is caused by mutations in the genes coding for keratins type 5 and 14, which form an important heterodimer for the formation of the keratin intermediate filament.

In order to simulate the proteins using bioinformatics, it was necessary to search for the gene sequences of the *KRT5* and *KRT14* genes and select a compatible target model to simulate the mutations that cause the main subtypes of Epidermolysis Bullosa Simplex (EBS). After observing the changes in the three-dimensional structures of K5 and K14 using the SWISS-MODEL, it was possible to demonstrate that mutations occur at specific sites in the genes, which are responsible for a subtype of EBS, and that intermolecular and intramolecular interactions are important for maintaining the three-dimensional structure and preserving the normal function of the proteins.

The prediction of the structures and functions of proteins can be used to design drugs targeted against protein alterations using bioinformatics programs, as they are able to simulate the interactions between the drug and the therapeutic target. As this is a relatively new area, it has the potential to be applied in various areas focused on biochemistry and molecular biology, and in the health area, the main objective is to improve the quality of life of patients with genetic diseases or pathologies caused by physical, chemical and biological agents.

Bibliographical references

ABBAS, A. K; LICHTMAN, A. H. Innate immunity, The initial defense against infection. In: ABBAS, A. K; LICHTMAN, A. H. **Basic Immunology**. 3. ed. Rio de Janeiro: Elsevier, 2009. p. 23-43.

ALMEIDA JR, H. L. Molecular Genetics of Bullous Epidermolysis. **Anais Brasileiros de Dermatologia**, Rio de Janeiro, v. 77, n. 5, p. 519-532, Sept./Oct. 2002.

ALMEIDA JR, H. L. Cytokeratins. **Anais Brasileiros de Dermatologia**, Rio de Janeiro, v.79, n.2, p. 135-145, mar./abr. 2004.

ARNOLD, K. et al. The SWISS-MODEL workspace: a web-based environment for protein structure homology modelling. **Bioinformatics (Oxford, England)**, Oxford, v. 15-22, n. 2, p. 195-201, jan. 2006.

ARIN, M. J. et al. Identification of novel and known KRT5 and KRT14 mutations in 53 patients with epidermolysis bullosa simplex: correlation between genotype and phenotype. **The British journal of dermatology**, Oxford, v. 162, n. 6, p. 1365-9, jun. 2010.

BIASINI, M. et al. SWISS-MODEL: modelling protein tertiary and quaternary structure using evolutionary information. **Nucleic Acids Research**, Oxford, v. 42 n. (W1): W252-W258, Apr. 2014.

BOEIRA, V. L. S. Y. **Hereditary Epidermolysis Bullosa: A Literature Review**. 2012. 42 f. Medical Course Conclusion Paper, Bahia Medical School, Federal University of Bahia, Salvador, 2012.

BOLLING, M. C. et al. Mutations in KRT5 and KRT14 cause epidermolysis bullosa simplex in 75% of the patients. **British Journal of Dermatology**, Oxford, v. 164, n. 3, p. 637-644, mar. 2011.

BRAGULLA, H. H; HOMBERGER, D. G. Structure and functions of keratin proteins in simple, stratified, keratinized and cornified epithelia. **Journal of anatomy**, London, v. 214, n. 4, p. 516-59, Apr. 2009.

BCHETNIA, M. et al. Expression signature of epidermolysis bullosa simplex. **Human genetics**, Berlin, v. 131, n. 3, p. 393-406, mar. 2012.

CHEN, H. et al. Keratin 14 gene mutations in patients with epidermolysis bullosa simplex. **The Journal of investigative dermatology**, Baltimore, v. 105, n. 4, p. 629-32, Oct. 1995.

CIUBOTARU, D. et al. Epidermolysis bullosa simplex in Israel: clinical and genetic features. **Archives of dermatology**, Chicago, v. 139, n. 4, p. 498-505, apr. 2003.

COOPER, M. P. **The cell**: A molecular approach. Available at:

<http://www.ncbi.nlm.nih.gov/books/NBK9839/> accessed on: May 20, 2016.

DIAS, S. F; TAVEIRA, C. C. Protein homology as a tool in the construction of new drugs. **Cenarium Pharmacêutico**, Brasilia, Year 4, No. 4, ISSN: 1984-33801, May/Nov 2011.

FINE, J. D. Inherited epidermolysis bullosa. **Orphanet Journal of Rare Diseases**, Paris, v. 5, n. 12, p. 1-17, May 2010.

FINE, J. D. et al. Inherited epidermolysis bullosa: Updated recommendations on diagnosis and classification. **Journal of the American Academy of Dermatology**, St. Louis, v. 70, n. 6, p. 1103-26, jun. 2014.

GARCIA, M. et al. Two novel recessive mutations in KRT14 identified in a cohort of 21 Spanish families with epidermolysis bullosa simplex. **The British journal of dermatology,** England, v. 165, n. 3, p. 683-92, sep. 2011.

GAO, Y. et al. A novel large deletion mutation of FERMT1 gene in a Chinese patient with Kindler syndrome. **Journal of Zhejiang University. Science. B**, Hangzhou, v. 16, n. 11, p. 957-62, nov. 2015.

GHR - Genetics HomeReference . **KRT5**. 2016. Available at :<

http://ghr.nlm.nih.gov/gene/KRT5>. Accessed on May 20, 2016a.

GHR - Genetics HomeReference . **KRT14**. 2016. Available at :<

http://ghr.nlm.nih.gov/gene/KRT14>. Accessed on May 20, 2016b.

GHR - Genetics Home Reference. **Junctional epidermolysis bullosa**. 2015. Available at: < http://ghr.nlm.nih.gov/condition/junctional-epidermolysis-bullosa> Accessed on October 26, 2015a.

GHR - Genetics Home Reference. **Dystrophic epidermolysis bullosa**. 2015. Available at: < http://ghr.nlm.nih.gov/condition/dystrophic-epidermolysis-bullosa> Accessed on October 26, 2015b.

JERABKOVA, B. et al. Keratin mutations in patients with epidermolysis bullosa simplex: correlations between phenotype severity and disturbance of intermediate filament molecular structure. **The British journal of dermatology**, England, v. 165, n. 5, p. 1004-13, May. 2010.

JUNQUEIRA, L. C; CARNEIRO, J. The cytoplasm. In: JUNQUEIRA, L.C; CARNEIRO, J. **Histologia Bàsica**. 10. ed. Rio de Janeiro: Guanabara, 2004. p. 23-51.

JOINTS. **Cellular junctions**. 2016. Available at: <

http://images.slideplayer.com.br/1/340412/slides/slide_44.jpg, > accessed on May 21, 2016.

KANG, T. W. et al. Novel and recurrent mutations in Keratin 5 and 14 in Korean patients with

Epidermolysis bullosa simplex. **Journal of dermatological science**, Amsterdam, v. 57, n. 2, p. 90-4, feb. 2010.

KEMP, M. W. et al. A novel deletion mutation in keratin 5 causing the removal of 5 amino acids and elevated mutant mRNA levels in Dowling-Meara epidermolysis bullosa simplex. **The Journal of investigative dermatology**, Baltimore, v. 124, n. 5, p. 1083-5, May. 2005.

LANE, E. B; MCLEAN, W. H. I. Keratins and skin disorders. **Journal of Pathology**, London, v. 204, n. 4, p. 355-66, nov. 2004.

LEE, C. H. et al. Structural basis for heteromeric assembly and perinuclear organization of keratin filaments. **Nature structural & molecular biology**, New York, v. 17;19, n. 7, p. 707-15, jun. 2012.

LIOVIC, M. et al. A mutation (N177S) in the structurally conserved helix initiation peptide motif of keratin 5 causes a mild EBS phenotype. **Experimental Dermatology**, Copenhagen, v.13, n.5, p. 332-334, mai. 2004.

MACHADO, A. B. et al. Epidermolytic palmoplantar hyperkeratosis (Vomer) Case report and literature review. **Anais Brasileiros de Dermatologia**, Rio de Janeiro, v. 77, n. 5, p. 593-603, Sep/Oct 2002.

MASON, J. D; ARNDT, K. M. Coiled Coil Domains: Stability, Specificity, and Biological Implications. **Chembiochem: a European journal of chemical biology**, Weinheim, v. 65, n. 2, p. 170-6, feb. 2004.

MASON, J. D; ARNDT, K. M. iPEP: peptides designed and selected for interfering with protein interaction and function. **Biochemical Society transactions**, London, v. 36, n. 6, p. 1442-7, dec. 2008.

MINAKAWA, S. et al. Mutational analysis on 16 Japanese population cases with epidermolysis bullosa simplex. **Journal of dermatological science**, Armsterdam, v. 72, n. 3, p. 330-2, dec. 2013.

MÜLLER, F. B. et al. Novel and recurrent mutations in keratin KRT5 and KRT14 genes in epidermolysis bullosa simplex: implications for disease phenotype and keratin filament assembly. **Human mutation**, New York, v. 27, n. 7, p. 719-20, jul. 2006.

OLIVA, G. Bioinformatics: Perspectives in Medicine. **Gazeta Mèdica da Bahia**, Bahia, v. 78, n. 1, p. 52-58, jan. 2008.

OLDAK, M. et al. Novel sporadic and recurrent mutations in KRT5 and KRT14 genes in Polish epidermolysis bullosa simplex patients: further insights into epidemiology and genotype-phenotype correlation. **Journal of Applied Genetics**, Pozan, v. 57, n. 1, p. 175181, jan. 2016.

PACHECO, A. G. M. **Comparative molecular modeling and molecular docking studies of the**

enzymalanosterol 14A-desmethylase from Moniliophthora perniciosa. 2009. 93 f. Dissertation (post-graduation), Biotechnology Post-Graduation Program, State University of Feira de Santana, Feira de Santana, 2009.

PROTEiNAS. **Intermediate Filament Proteins**. 2016. Available at: <http://www.lookfordiagnosis.com/mesh_info.php?term=Prote%C3%ADnas+de+Filamentos +Intermediarios&lang=2> access on: May 20, 2016.

RUGG, E. L. et al. Epidermolysis Bullosa Simplex in Scotland Caused by a Spectrum of Keratin Mutations. **Journal of Investigative dermatology**, Baltmore, v. 127, n. 3, p. 547-80 mar. 2007.

SANTOS FILHO, O. A; ALENCASTRO, R. B. MODELING OF PROTEINS BY HOMOLOGY, **Quimica Nova**, Sao Paulo, v. 26 n. 2, p. 253-259, Mar./ Apr. 2003.

SCHUILENGA-HUT, P. H. et al. Mutation analysis of the entire keratin 5 and 14 genes in patients with epidermolysis bullosa simplex and identification of novel mutations. **Human mutation**, New York, v. 21, n. 4, p. 477, abr. 2003.

SEIDEL, E. J. et al., Comparison between the Ward method and the K-means method in the grouping of milk producers. **Science and Nature**. Santa Maria, v. 30, n. 1, p. 7-15, jan. 2008.

SOUZA, M. L. **Homology modeling of protein structures from Apis mellifera venom**. 2008. 72f. Dissertation (Graduation) Biomedical Informatics University of Sao Paulo, Ribeirao Petro, 2008.

S0RENSEN, C. B. et al. Identification of novel and known mutations in the genes for keratin 5 and 14 in Danish patients with epidermolysis bullosa simplex: correlation between genotype and phenotype. **The Journal of investigative dermatology**, Baltimore, v. 112, n. 2, p. 18490, feb. 1999.

SZEVERENYI, I. et al. The Human Intermediate Filament Database: Comprehensive Information on a Gene Family Involved in Many Human Diseases. **Human mutation**, New York, v. 29, n. 3, p. 351-60, mar. 2008.

TITEUX, M. et al. Three Severe Cases of EBS Dowling-Meara Caused by Missense and Frameshift Mutations in the Keratin 14 Gene. **Journal of Investigative Dermatology**, Baltimore, v. 126, n. 4, p. 773-6, Apr. 2006.

VERLI, H. **Bioinformatics: from Biology to Molecular Flexibility**. 1. ed. Sao Paulo: SBBq, 2014.

YASUKAWA, K. et al. Epidermolysis bullosa simplex in Japanese and Korean patients: genetic studies in 19 cases. **The British journal of dermatology**, Oxford, v. 155, n. 2, p. 3137, aug. 2006.

ZHANG, Z. et al. A greedy algorithm for aligning DNA sequences. **Journal of computational biology: a journal of computational molecular cell biology**, New York, v. 7, n. 1-2, p. 203-14, feb./abr. 2000.

Molecular analysis of mutations associated with the *col7Ai* gene in DISTROPHIC BOLD EPIDERMOLYSIS (EBD)

Laura Lisieux Dos Santos Monteiro*

Paulo Roberto Martins Queiroz**

*Graduate student in Biomedicine at the University Center of Brasilia - UniCEUB.

**Biologist. PhD in Animal Biology from the University of Brasilia. Email: pqsilva@uol.com.br

Introduction

Epidermolysis Bullosa (EB) is a group of hereditary diseases characterized by defects in intraepidermal or dermoepidermal adhesion which result in the appearance of blisters on the skin in response to the slightest trauma (ALMEIDA JR, 2002). EB's fall into four types which are classified according to the histology and formation of the blisters: Simple EB, Junctional EB, Dystrophic EB and Kindler's Syndrome, a group recently included in the third international consensus on EB (BOEIRA, 2012).

With regard to the epidemiology of EB, various studies have been carried out to estimate the incidence and prevalence of the disease using different techniques, but the most important studies and those considered to be the benchmark are those derived from the National EB Registry in the United States (BOEIRA, 2012). According to the US, there are 50 cases of EB per one million live births, 92% of which are simple EB, 5% dystrophic, 1% junctional and 2% unclassified. There is no epidemiological data in Brazil, as studies are still being carried out on the number of individuals affected by Epidermolysis Bullosa. EB affects both genders and can occur in all ethnic groups (GÜRTLER; DINIZ; FILHO, 2005).

Epidermolysis bullosa dystrophica (EBD) is the least common and most severe type of EB. It is subdivided into four main subtypes, two in the autosomal dominant form and two in the autosomal recessive form (OLIVEIRA, et al., 2010). The subtypes of dominant EBD (EBDD) are: *Cockayne-Touraine* (EBDD-CT), whose clinical characteristics are the formation of blisters on the extremities of the body, oral cavity (mucous membranes and teeth) and dystrophy or absence of nails; and the *Pasini* subtype (EBDD-Pa), with the same clinical manifestations as EBDD-CT, but in a more severe form, with the formation of more extensive blisters, similar to papules, also appearing on the trunk region. The subtypes of recessive EBD (EBDR) tend to be more aggressive (ALMEIDA JR, 2002). The recessive *non-Hallopeau-Siemens* form (EBDR-nHS) is more severe than the dominant subtypes and is also clinically characterized by the development of blisters on the extremities of the body and loss of nails, but with little involvement of the mucous membranes. The recessive *Hallopeau-Siemens* form (EBDR-HS) is the most severe of the EBD group. Its clinical manifestations include the formation of blisters with generalized distribution throughout the body; pseudosyndactyly (deformity of the hands and feet in the shape of a "box glove"), extremely affected nails and teeth with loss; involvement of the internal mucous membranes with esophageal restrictions, anal and urethral stenosis (narrowing); phimosis; corneal bedsores; iron deficiency anemia and developmental deficits caused by malnutrition. In addition, these patients are at serious risk of developing skin cancer at the sites of chronic lesions (BARROS; RASKIN; PEREIRA-FERRARI, 2004; BOEIRA, 2012).

All subtypes of EBD, dominant and recessive, come from mutations in the same gene, *COL7A1*.

However, what determines each subtype is the location and type of mutation that occurs in this gene. *COL7A1* encodes type VII collagen, an important protein that makes up the anchoring fibrils that participate in the adhesion between dense laminin and the dermis in the skin (DANG; MURRELL, 2008).

More than 730 mutations have been described for EBD, leading to different disease phenotypes. Among these mutations, the most common types are: mutations that code for a different amino acid (*missense*); *nonsense* mutations, which code for a stop codon (*nonsense*); deletions; insertions; *indel* mutations, when deletions and insertions occur concomitantly; and those that occur in *splicing* (DANG; MURRELL, 2008; LIN et al., 2012).

The diagnosis of EB is made on the basis of the patient's clinical and laboratory findings. In the clinic, the family history is assessed and whether there is any consanguinity between the parents. In the laboratory, the diagnosis is made using electron microscopy (EM), the gold standard method, which consists of the histological analysis of a fragment of skin tissue. Treatment involves caring for the patient's wounds and blisters using sterile dressings to avoid possible infections. Nutritional support is provided and topical or systemic antibiotics are used when contamination occurs. Preventing the appearance of blisters is the main aim of treatment, because if they become chronic lesions, they can develop into malignant tumors, one of the main causes of death in patients with EBD (BOEIRA, 2012).

The biomolecular study of bullous epidermolysis in Brazil is still insufficient, and it is necessary to resort to international studies in other languages. A more complete study is needed on these parameters because, in addition to increasing the number of national studies in this field, the specific genetic study allied to molecular biology is of great help in the development of gene therapies. According to Cutlar, Greiser and Wang (2014), the replacement of a defective gene with a fully functional copy is the basic principle of gene therapy. The patient starts to produce the protein - which was previously synthesized in a defective way - correctly, reducing the symptoms of the disease and improving their quality of life considerably.

The aim of this study was to associate the main types of mutations with the molecular structure of the COL7A1 protein related to the phenotypes of Epidermolysis Bullosa Dystrophica (EBD).

Methodology

In order to achieve the proposed objective, an experimental study was carried out in three stages, consisting of a bibliographical review of scientific articles on Dystrophic Epidermolysis Bullosa, processing of the information obtained using bioinformatics programs and preparation of the results.

For the literature review, scientific articles and theses were searched in the *EBSCOhost*, Scielo, Medline, PubMed, Elsevier and Bireme databases, using the keywords "dystrophic epidermolysis bullosa", "*COL7A1*", "mutations and *COL7A1*", "dystrophic epidermolysis bullosa" and "mutations and dystrophic epidermolysis bullosa", and the search period was defined as between 2002 and 2016, not excluding previous works of importance to better substantiate the topic. Criteria were also defined as papers written in Portuguese and English.

Bioinformatics

Bioinformatics refers to the use of computational tools in the study of biological problems and issues, also covering applications related to human health such as the planning of new drugs (VERLI, 2014).

Bioinformatics was the main part of the work, as it was used to carry out sequence and structural analyses using the programs *The Sequence Manipulation Suite* (http://www.bioinformatics.org/sms2/), *PDB - Protein Data Bank* (http://www.rcsb.org/pdb/home/home.do), *Swiss Model* (http://swissmodel.expasy.org/) and *Phyre 2* (http://www.sbg.bio.ic.ac.uk/~phyre2/html/page.cgi?id=index). All analysis of the *COL7A1* gene was carried out using the *NCBI - National Center for Biotechnology Information* database (http://www.ncbi.nlm.nih.gov/).

Firstly, the DNA sequence of the *COL7A1* gene was searched for in the *NCBI* database and then the coding region (CDS) (ID 1294), which refers to the protein sequence of type VII collagen, was analyzed. To obtain it, we searched *NCBI* for the *COL7A1* gene transcript (mRNA) and, when we chose the option "Homo sapiens collagen type VII alpha 1 (COL7A1), mRNA", just below the scientific references, the CDS appeared as a clickable option. After clicking, the coding region is highlighted for viewing. With this region, attempts were made to model the type VII collagen protein chain in the *Swiss Model* and *Phyre 2* programs without success, as it was not in the databases. It was therefore decided to carry out a new search in which the regions corresponding to the mutations to be studied were determined and modeled as a single artificial structure, but the modeling programs found no data, meaning that it was not possible to model the sequences.

To arrive at the final sequence used in this work, the protein segment of type VII collagen was re-entered into the *Swiss Model* program for another attempt. From the similar chains that were suggested, a variant was selected that is in line with the structures that are described in the scientific articles researched that form the basis of this work. The segment was then modeled in *Swiss Model* and its structure provided a simulation of protein chains under pathological conditions, i.e. with the types of mutations that correspond to the phenotypes of Dystrophic Epidermolysis Bullosa.

Results and Discussion

Dystrophic epidermolysis bullosa and the *COL7A1* gene.

Table 1: Clinical characteristics of EBD subtypes.

SUBTYPES	CLINICAL ASPECTS	
Epidermolysis Bullosa Dominant Dystrophic *Cockayne-Touraine* (EBDD-CT)	Formation of blisters on the extremities of the body and oral cavity (mucous membranes and teeth). Dystrophy or absence of nails.	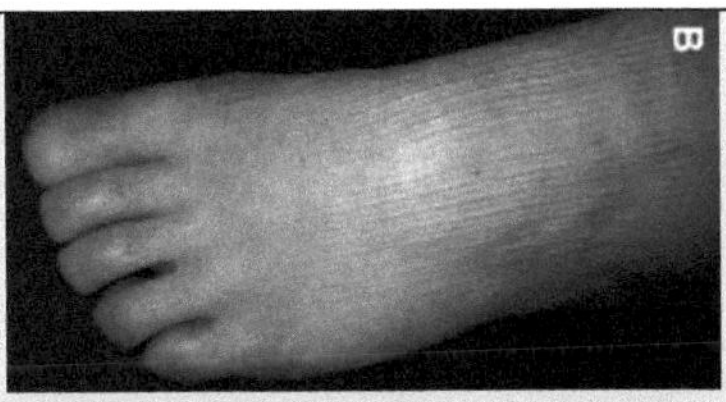 B
Epidermolysis Bullosa Dominans *Pasini* (EBDD-Pa)	Formation of more extensive blisters, on the extremities of the body, chest and oral cavity cavity (mucous membranes and teeth). Dystrophy or absence of nails.	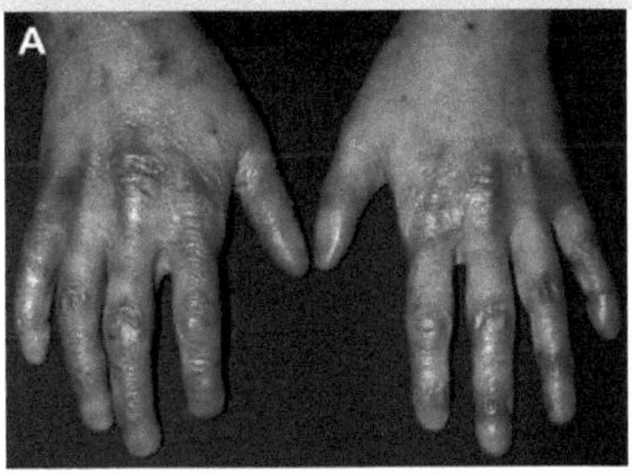 A
Epidermolysis Bullosa Recessiva non *Hallopeau-Siemens* (EBDR-nHS)	Formation of blisters on the extremities of the body, loss of nails. More severe than the dominant forms. Little involvement of mucous membranes.	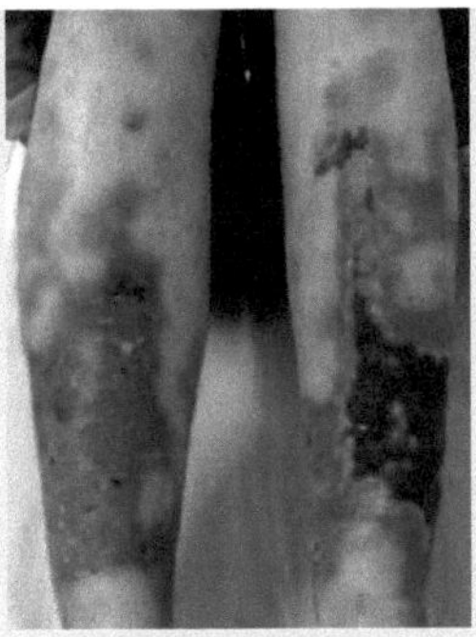 Source: GÜRTLER; DINIZ; FILHO (2005).

Epidermolysis Bullosa Recessiva *Hallopeau-Siemens* (EBDR-HS)	More severe subtype. Blisters with generalized distribution over the body; pseudosyndactyly; loss of nails and teeth; compromised internal mucous membranes;	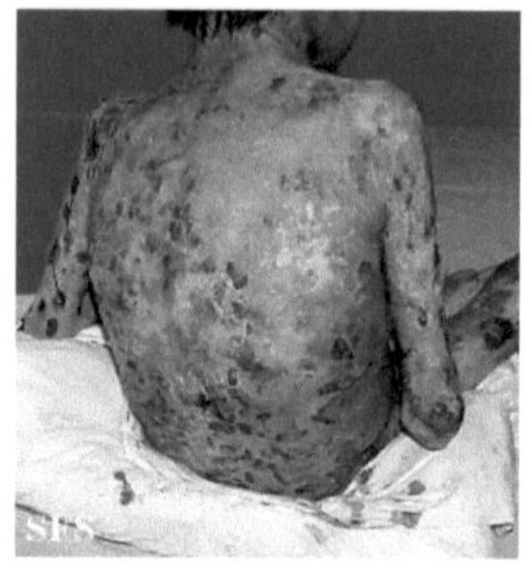

Source: Adapted from BOEIRA (2012).

Epidermolysis bullosa dystrophica (EBD) is a hereditary condition divided into subtypes characterized by the distribution of blisters along the patient's skin and mucous membranes (BARROS; RASKIN; PEREIRA-FERRARI, 2004; BOEIRA, 2012). Table 1 shows the four subtypes of EBD and their clinical characteristics.

Figure 1: Ultrastructure of the basement membrane zone and proteins involved in EB.

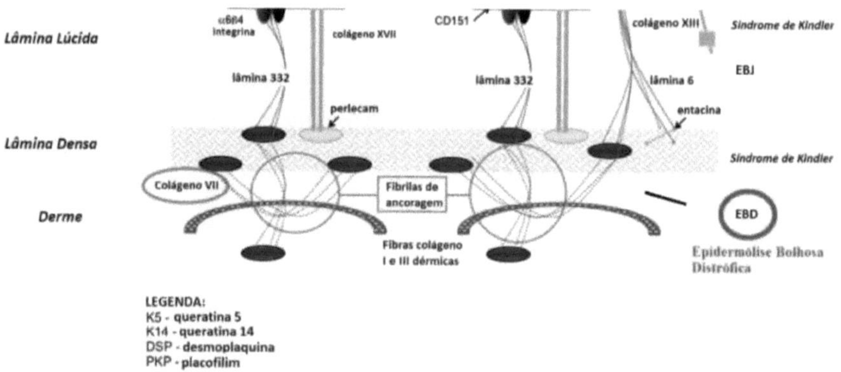

Source: Adapted from BOEIRA (2012).

The defects that cause EBD phenotypes are found in the type VII collagen protein that forms the anchoring fibrils present between the dense lamina and the dense sublamina at the dermal-epidermal junction, where blisters are formed, as shown in Figure 1 (DANG; MURRELL, 2008).

The four main subtypes of EBD described are the result of mutations in the same gene, *COL7A1*, which codes for type VII collagen and is present on chromosome 3 at position p21.31 (Figure 2) (WEINEL, 2008). Each subtype varies according to the type of mutation and its location, from the most severe phenotype, the recessive *Hallopeau-Siemens* form, to the mildest, the dominant *Cockayne-Touraine* and *Pasini* forms (BOEIRA, 2012).

Figure 2: Physical mapping of the *COL7A1* gene (GenBank ID 1294) at *locus* 3p21.31.

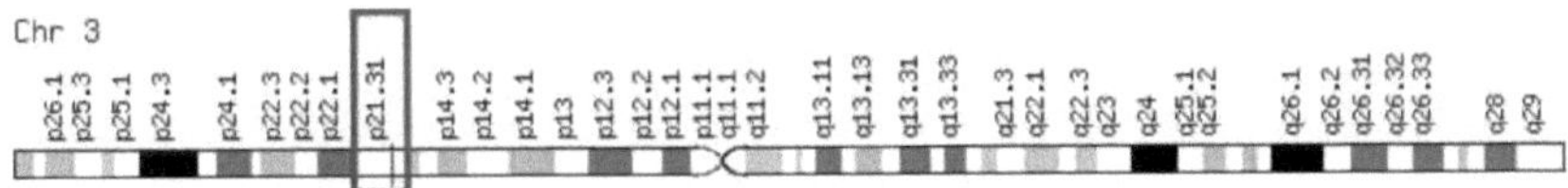

Source: Adapted from GENECARDS (2016).

1.2. Type VII collagen

Type VII collagen is produced by epidermal keratinocytes and dermal fibroblasts. During intracellular processing, the mRNA transcript is translated into a proα1 (VII) polypeptide chain, which joins with two other proα1 chains, generating a 2944 amino acid structure consisting of a central triple-helical collagen domain preceded and followed by two non-collagenous domains, NC-1 (non-collagenous-1), a large amino-terminal chain, and the smaller NC-2 (non-collagenous-2), a carboxy-terminal portion, as can be seen in the simplified representation in Figure 3 (DANG; MURRELL, 2008).

Figura 3: Schematic representation of the type VII collagen molecule.

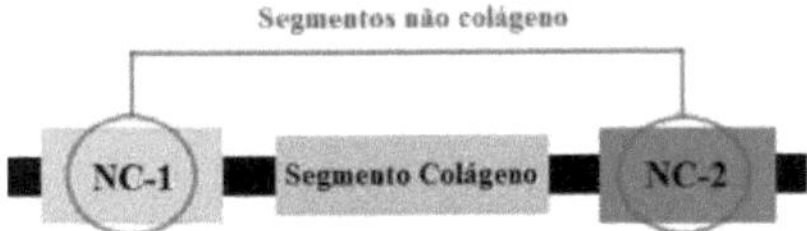

Source: Adapted from ALMEIDA JR (2012).

The center of its structure also has a small non-collagenous segment, which gives the protein flexibility. The fusion of two triple helical molecules occurs at the extracellular level, through the loss of the NC-2 segment, resulting in antiparallel dimers whose union, in turn, will form the anchoring fibrils (Figure 4) (ALMEIDA JR, 2002; VARKI et al., 2006).

Figura 4: Formation of type VII collagen under physiological conditions (left side) and pathological conditions (right side), showing the subtypes of EBD.

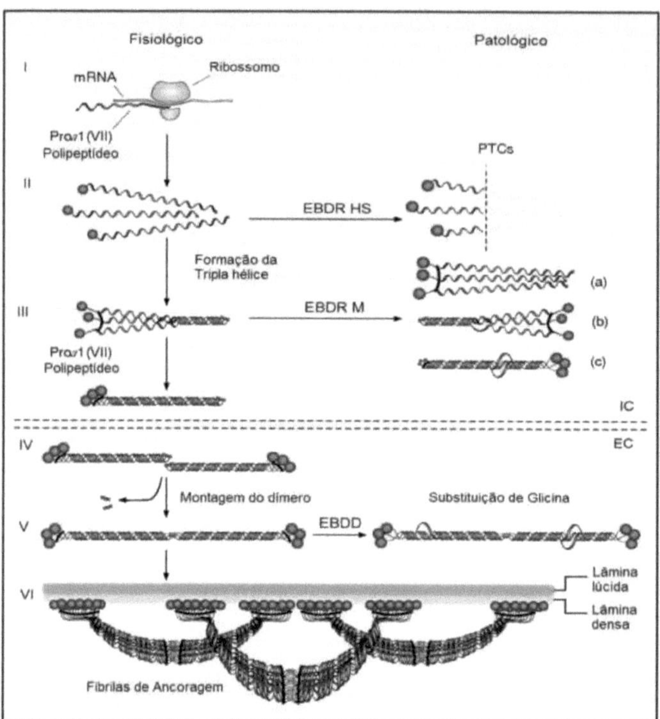

Source: Adapted from VARKI et al. (2006).

It is an extremely important protein in the dermal-epidermal junction, as it provides firmness and flexibility to the anchoring fibrils, thus keeping the basement membrane zone (BMZ) and the dermis together and firm. If the collagen produced is defective, the maintenance of this junction is impaired. The result is that blisters form with every friction the skin suffers, no matter how small.

As previously mentioned, more than 730 mutations have been described for EBD that result in defective collagen. The consequences vary according to the type of mutation, such as PTC *(Premature Termination Codon)*, shown in Figure 4, which is generally associated with *nonsense* mutations (coding for a stop amino acid) and deletions, which are frequent in the EBDR-HS phenotype when present in both alleles. Mutations coding for a different amino acid (*missense*), mainly referred to as substitutes for the amino acid glycine (Gly) in Dystrophic Epidermolysis Bullosa, and insertions are commonly associated with the EBDR-nHS phenotype and also with the dominant subtypes (EBDD-CT and EBDD-Pa), which are milder forms. These mutations can cause interference in the association of collagen chains and in the formation of the triple helix, thus destabilizing this molecule. Substitutions of the amino acid Gly (Glycine), also in the triple helix of collagen, cause interference in the formation of the dimers that make up the anchoring fibrils, common in EBDD phenotypes (ALMEIDA JR, 2002; VARKI et al., 2006; DANG; MURRELL, 2008).

The arrangement of the triple helix of type VII collagen, the region of study in this work, is made up of repeated sessions of the Gly-X-Y motif, where X and Y can be any amino acid, but generally their positions are occupied by Proline (PRO) and Hydroxyproline (HYP) (Figure 5) (MANKA et al., 2012).

Figura 5: The composition of the collagen molecule.

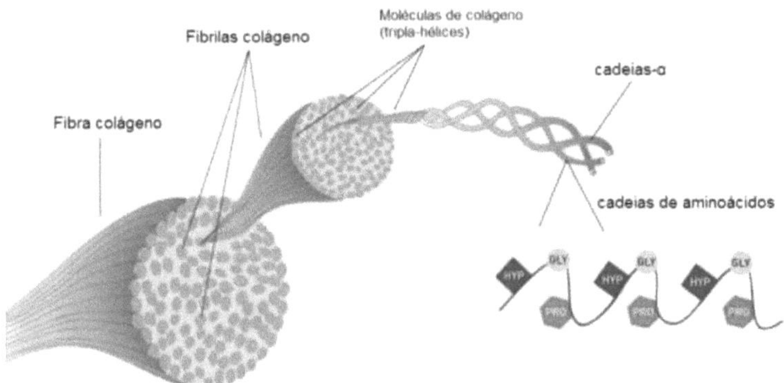

Source: PROTO-COL UK (2014).

The helical structure of collagen's α-chains (proα1) gives the protein the rigidity it needs to hold the dermis and epidermis together. This is because each amino acid in this repeat has a precise function: the glycine side chain that corresponds to a hydrogen atom is the only portion that can fit into the tight center of the triple helix structure. Hydrogen bridges are favored by the chemical groups present in the peptide bond between the NH portion of a glycine residue and a carbonyl group (C=O), thus holding the three chains together. The fixed C-N peptide-proline or peptide-hydroxyproline angle allows the polypeptide chain, when folded into a helix, to form a geometry in which the polypeptide chains can twist together to form a triple-helical chain, as shown in Figure 6. Interestingly, although the rigid peptide-proline bonds disrupt the amino acid structure in an α-helix chain, they stabilize the final triple-helix structure (LI; BRODSKY; BAUM, 2007; SHOULDERS; RINES, 2009).

Figura 6: Overview of the molecular structure of the collagen triple helix. In (a), the high-resolution crystal structure of the triple helix. In (b), the triple helix in three representations: *space-filling*, *ball-and-stick* and ribbon. In (c), a *ball-and-stick* segment of the triple helix with the interchain hydrogen bonds highlighted. And in (d), polypeptide bonds between the α-chains.

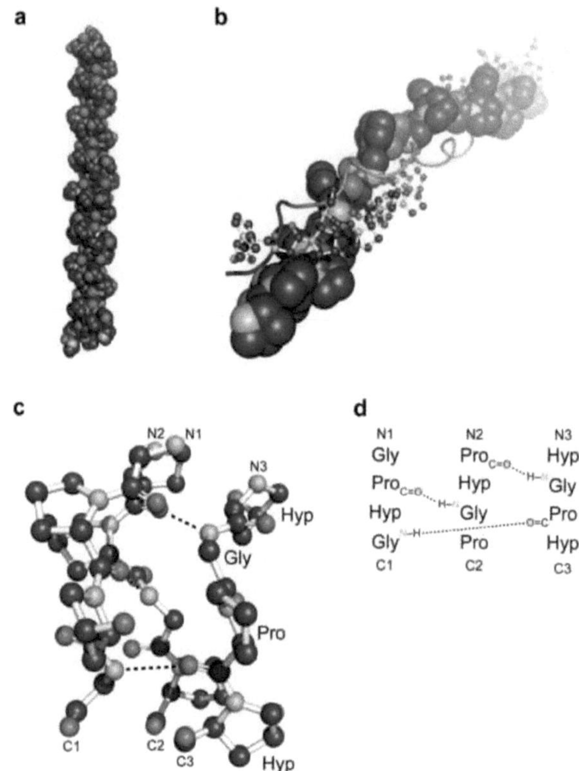

Source: SHOULDERS; RAINES (2009).

Experimental phase: modeling protein chains

The literature does not describe mutations that occur frequently in EBD. Therefore, there is no main mutation to be studied. For this reason, it was decided to study the types of mutations that lead to the different phenotypic levels of Dystrophic Epidermolysis Bullosa.

It is known that collagen protein chains change according to the mutation that occurs in the amino acid sequence and its location. These different structures were represented and analyzed in this work, comparing them to a normal chain, i.e. one without alterations or mutations.

The protein modeling program *Swiss Model* suggested the protein variant ID 4AUO.2.B: a 38 amino acid polypeptide with the Gly-X-Y motif that appears repeatedly in the structure (GPPGPPGPPGPQGLAGQRGIVGLPGQRGERGPPGPPGP) which corresponds to the B chain of the triple helix of the interstitial collagen protein present in type I, II and III collagens. Its constitution closely resembles that of type VII collagen and was therefore used to demonstrate the mutations related to EBD phenotypes (Figure 7) (BISIANI et al., 2014).

Figura 7: Representation of the normal B chain of the interstitial collagen protein.

Source: BIASINI et al. (2014).

The most common types of mutations recorded in the literature that lead to EBD are *missense, nonsense*, deletions, insertions and *indel* mutations. All of these types are shown in the following figures, with the exception of the *nonsense* mutation type, which generates a chain of amino acids that is insufficient to be modeled by the *Swiss Model*. These mutations cause a premature termination codon (PTC) resulting in a non-functional truncated protein which, in most cases, is related to the recessive phenotypes of EBD, especially the most severe form: *Hallopeau-Siemens*. This can occur in heterozygosity, PTC combined with another type of mutation such as PTC and a glycine substitution, and in homozygosity, when PTC occurs in both alleles (DANG; MURRELL, 2008).

Missense mutations code for a different amino acid. In EBD, they are usually associated with glycine substitutions resulting, in most cases, in the dominant phenotypes (EBDD-CT and EBDD-Pa), which are characterized by being clinically milder. This is because these substitutions cause only a small irregular twist at the site of the mutation in the protein's amino acid sequence, destabilizing its structure and compromising it, causing interference in the formation of the dimers that make up the anchoring fibrils (DANG; MURRELL, 2008; DENG et al., 2008; VARKI et al., 2006). Figure 8 shows this abnormal twisting.

Figura 8: Representation of a *missense* mutation circled in red: substitution of Glycine for Valine at position 10.

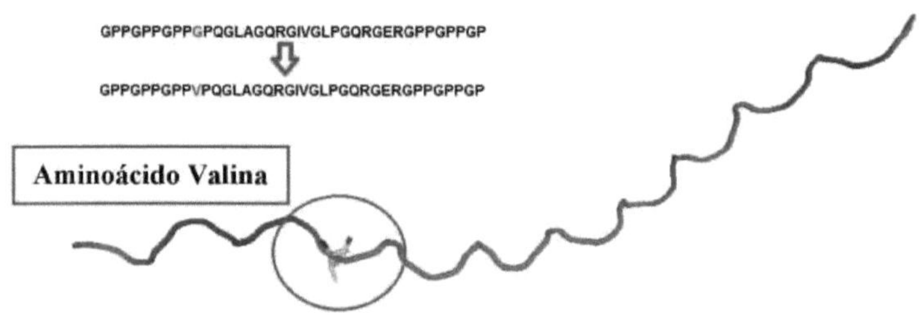

Source: Adapted from BIASINI et al. (2014).

Deletion mutations are related to recessive EBD phenotypes and, more often than not, to the most severe cases: *Hallopeau-Siemens* recessive EBD. Deletions cause very problematic alterations in the collagen protein, compromising its functionality almost completely. Figure 9 shows a deletion of the amino acid glycine which generated a motif with a pleated β-sheet structure (represented by the arrow shape) between the helix of the chain. It is known that the physical characteristic of collagen is its rigidity, which is provided by the helical structure of its protein chain (DANG; MURRELL, 2008; VARKI et al., 2006). The pleated β-sheet compromises this rigidity by making collagen too malleable, causing it to lose its real function of adherence and firmness to the skin, leaving it literally floppy.

Figura 9: Representation of a deletion-type mutation circled in red: Glycine deletion at position 16.

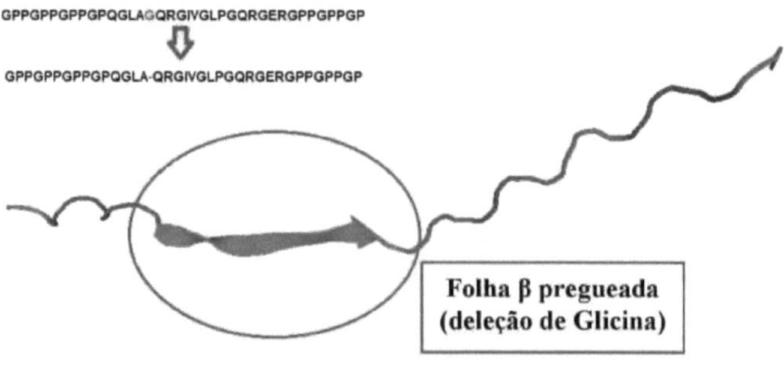

Source: Adapted from BIASINI et al. (2014).

Insertion mutations and *indel-type* mutations are related, as are deletions, most often to recessive phenotypes, especially EBDR-HP. The insertion of another amino acid into the protein chain causes interference between the repeats of the Gly-X-Y motif, changing the conformation of the protein and leading to a very altered twist that destabilizes its entire structure. This abnormality can be seen in Figure 10 and Figure 11 (B). The literature describes that *indel* insertions and mutations lead to PTC in the polypeptide chain, interfering with the entire functionality of collagen, thus resulting in the recessive phenotypes of EBD. As previously mentioned, the pleated β-sheet structure in the collagen chain, seen in Figure 11 (A), causes significant changes in the functionality of the protein, since its rigidity is extremely compromised. For this reason, *indel* mutations in both alleles will almost always be related to the more severe EBDR-HS phenotype (DANG; MURRELL, 2008).

Figura 10: Representation of an insertion-type mutation circled in red: insertion of Asparagine at position 17.

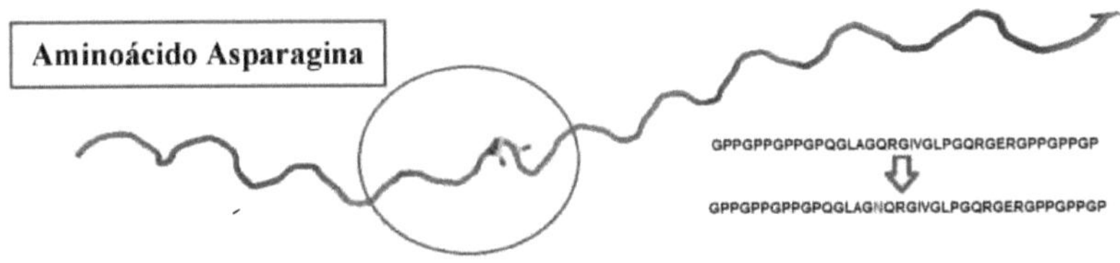

Source: Adapted from BIASINI et al. (2014).

Figura 11: Representation of an *indel-type* mutation: in **A**, deletion of Glycine at position 16 and in **B**, insertion of Asparagine at position 23.

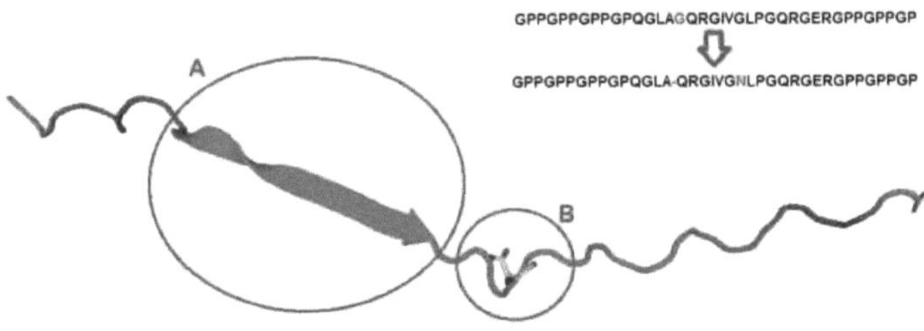

Source: Adapted from BIASINI et al. (2014).

Experiments have shown that *missense* mutations are generally related to autosomal dominant phenotype profiles: EBDD-CT and EBDD-Pa. In most cases, these are substitutions of the amino acid glycine and do not lead to complete loss of function of the type VII collagen molecule, causing clinically milder EBD phenotypes. The other types (*nonsense*, deletions, insertions and *indel)* are more associated with autosomal recessive phenotypes: EBDR-nHP and EBDR-HP. Mutations that cause these more severe clinical conditions result in a non-functional truncated protein, especially when *nonsense* mutations occur, as they cause a premature termination codon (PTC). They can also bring other conformations to the protein, such as the pleated β-sheet and abnormal twists, whose structures cause destabilization of the protein domain, impairing almost all of the functionality of collagen in the tissue.

Molecular studies related to Dystrophic Epidermolysis Bullosa are necessary, above all, in order to bring new treatment prospects, new drugs and faster and more advanced diagnostics to patients with this disease. A good example are gene therapies involving the use of recombinant DNA, which are currently being extensively studied and experimented with in order to considerably improve the quality of life of patients diagnosed with EBD (CUTLAR; GREISER; WANG, 2014). This requires prior knowledge of the molecular aspects of the individual's phenotype, such as the type of mutation and the level of functionality of type VII collagen in the tissue, as shown in the studies presented in this paper. In addition to these therapies, it is essential to know the functional aspects of type VII collagen. A recent study carried out at Oxford University showed that this type of collagen is associated with the development of squamous cell carcinoma (SCC), as it suppresses TNFβ signaling and angiogenesis, inhibiting tumor growth. In these cases, type VII collagen is being studied as a therapy for SCC (MARTINS et al., 2016). For these reasons, molecular studies are so important, as they help researchers search for new strategies to improve the quality of life of patients not only with EBD, but other genetic diseases with no cure.

Final considerations

According to the studies carried out, the phenotypes of Dystrophic Epidermolysis Bullosa differ according to the type of mutation that occurs in the gene sequence and also its location. Each of these types results in a protein chain with characteristics unfavorable to the functionality of type VII collagen, leading to a different phenotypic profile, varying clinically from mild to severe. The most common types of mutations are *missense*, *nonsense*, *indel* mutations, deletions and insertions.

It was concluded that the mutation that would cause the least clinical involvement would be the *missense* type and that Dystrophic Epidermolysis Bullosa can present clinically at various levels depending on the mutation that occurs in the *COL7A1* gene and that type VII collagen can be partially or completely non-functional.

Bibliographical references

ALMEIDA JR, H. L. Molecular Genetics of Epidermolysis Bullosa. **Anais Brasileiros de Dermatologia**, Rio de Janeiro, v. 77, n. 5, p. 519-532, Sept./Oct. 2002.

BARROS, I. R.; RASKIN, S.; PEREIRA-FERRARI, L. Dystrophic Epidermolysis Bullosa: Report of a Brazilian case. **Estudos de Biologia**, Curitiba, v. 26, n. 57, p. 25-30, Oct/Dec. 2004.

BIASINI, M.; BIENERT, S.; WATERHOUSE, A.; ARNOLD, K.; STUDER, G.; SCHMIDT, T.; KIEFER, F.; CASSARINO, T. G.; BERTONI, M.; BORDOLI, L.; SCHWEDE, T. SWISS-MODEL: modelling protein tertiary and quaternary structure using evolutionary information. **Nucleic Acids Research**, London, v.42, p. W252-W258, jul. 2014.

BOEIRA, V. L. S. Y. **Hereditary bullous epidermolysis: A literature review**. 2012. 42 f. Thesis (Monografia) - Medicine, Federal University of Bahia, Salvador, 2012.

CUTLAR, L.; GREISER, U.; WANG, W. Gene therapy: pursuing restoration of dermal adhesion in recessive dystrophic epidermolysis bullosa. **Experimental Dermatology**, Sydney, v. 23, n. 1, p. 1-6, jan. 2014.

DANG, N.; MURRELL, D. F. Mutation analysis and characterization of *COL7A1* mutations in dystrophic epidermolysis bullosa. **Experimental Dermatology**, Sydney, v. 17, n. 7, p. 553568, jul. 2008.

DENG, W.; CHEN, S.; LU, C.; ZHOU, X.; CHEN, M.; LAI, W.; WANG, Y. A novel p.Gly1700Aspmutation in COL7A1 responsible for dominant dystrophic epidermolysis bullosa: More severe phenotype in female members of a Chinese Family. **Journal of Dermatological Science**, Singapore, v.49, n. 2, p. 166-169, feb. 2008.

GENECARDS. ***COL7A1* Gene (Protein Coding)**. Rehovot, 2016. Available at: <http://www.genecards.org/cgi-bin/carddisp.pl?gene=COL7A1>. Accessed on: June 22, 2016.

GÜRTLER, T. G. R.; DINIZ, L. M.; FILHO, J. B. S. Epidermolysis bullosa recessiva mitis - Clinical case report. **Anais Brasileiros de Dermatologia**, Rio de Janeiro, v. 80, n. 5, p. 503-508, Sep/Oct 2005.

LI, Y.; BRODSKY, B.; BAUM, J.; NMR Shows Hydrophobic Interactions Replace Glycine Packing in the Triple Helix at a Natural Break in the (GLY-X-Y) Repeat. **The Journal of Biologica Chemistry**, Rockville, v. 282, n. 31, p. 22699-22706, Aug. 2007.

LIN, Y.; CHEN, X. J.; LIU, W.; GONG, B.; XIE, J.; XIONG, J. H.; CHENG, J.; DUAN, X. L.; LIN, Z. C.; HUANG, L. L.; WAN, H. Y.; LIU, X. Q.; SONG, L. H.; YANG, Z. L. Two Novel Mutations on Exon 8 and Intron 65 of COL7A1 Gene in Two Chinese Brothers Result in Recessive Dystrophic

Epidermolysis Bullosa. **Journal Plos One**, Columbus, v. 7, n. 11, p. e50579, nov. 2012.

MANKA, S. W.; CARAFOLI, F.; VISSE, R.; BIHAN, D.; RAYNAL, N.; FARNDALE, R. W.; MURPHY, G.; ENGHILD, J. J.; HOHENESTER, E.; NAGASE, H. Structural insights into triple-helical collagen cleavage by matrix metalloproteinase 1. **PNAS Early Edition**. Washington.DC, v. 109, n. 31, p. 12461-12466, jul. 2012.

MARTINS, V. L.; CALEY, M. P.; MOORE, K.; SZENTPETERY, Z.; MARSH, S. T.; MURREL, D. F.; KIM, M. H.; AVARI, M.; MCGRATH, J. A.; CERIO, R.; KIVISAARI, A.; KĂHĂRI, V. M.; HODIVALA-DILKE, K.; BRENNAN, C. H.; CHEN, M.; MARSHAL, J. F.; O'TOOLE, E. A. Suppression of TGFβ and Angiogenesis by Type VII Collagen in Cutaneous SCC. **Journal of the National Cancer Institute**, London, v. 108, n. 1, p. 1-11, jan. 2016.

MEDICINAPERTUTTI. **Basal membrane.** Borgorosis, 2015. Available at: <http://medicinapertutti.altervista.org/argomento/membrana-basale/>. Accessed on: May 19, 2016.

OLIVEIRA, Z. N. P.; PÉRIGO, A. M.; FUKUMORI, Ligia M. I.; AOKI, V. Immunomapping in hereditary epidermolysis bullosa. **Anais Brasileiros de Dermatologia**, Rio de Janeiro, v. 85, n. 6, p. 856-861, nov/dez. 2010.

PROTO-COL UK. **What is collagen?** London, 2014. Available at: <http://www.proto-col.com/blog/2014/07/collagen/>. Accessed on: June 19, 2016.

SHOULDERS, M. D.; RAINES, R. T.; Collagen structure and stability. **Annual Reviews of Biochemistry**, Palo Alto, v. 78, n. 1, p. 929-958, mar. 2009.

VARKI, R.; SADOWSKI, S.; UITTO, J.; PFENDNER, E. Epidermolysis bullosa. II. Type VII collagen mutations and phenotype-genotype correlations in the dystrophic subtypes. **Journal of Medical Genetics**, London, vol. 44, n. 3, p. 181-192, Mar/Apr. 2006.

VERLI, H. **Bioinformatics: from Biology to Molecular Flexibility.** Sao Paulo: SBBq, 2014.

Printed by Books on Demand GmbH, Norderstedt / Germany